TEAM SCIENCE

TEAM SCIENCE

Organizing Classroom Experiments That Develop Group Skills

Marilyn Coffin

Zephyr
Press®

REACHING THEIR HIGHEST POTENTIAL
Tucson, Arizona

Team Science
Organizing Classroom Experiments That Develop Group Skills

Grades 5–8

© 1996 by Zephyr Press
Printed in the United States of America

ISBN 1-56976-013-6

Editors: Stacey Lynn and Stacey Shropshire
Cover design: Nancy Taylor
Design and production: Daniel Miedaner

Zephyr Press
P.O. Box 66006
Tucson, Arizona 85728-6006

Library of Congress Cataloging-in-Publication Data are available

To Chris, Lauren, Madison, and Shirley,
the best support team ever.

CONTENTS

APPENDIX A: Forms

APPENDIX B

INTRODUCTION

Organizing Labs

Key Words

Lab members must *cooperate, focus on the task,* and *work together.*

Basic Format

Each lesson has four parts:

1. Three-minute set-up
2. Five-minute warm-up
3. Twenty-five–minute experiment
4. Five-minute clean-up

Three-Minute Set-up

During the three-minute set-up, each member of the group is responsible for performing a specific task.

Recorder gets a pencil, paper, and the lab folder.

Controller clears and organizes the work area.

Hopper gets the lab tray with the materials.

Observer arranges the chairs around the work area.

Five-Minute Warm-up

The five-minute warm-up allows the teacher to review the safety rules that apply to the lab, review or demonstrate the techniques or skills required, and tell the hoppers where materials are that aren't in the lab trays. Finally, the teacher reminds all group members to follow along on the lab sheets as the recorder reads aloud the directions for the experiment.

Twenty-five–Minute Experiment

The twenty-five–minute experiment is very structured. Again, each member of the group has specific tasks to perform. The tasks rotate with each new experiment. However, the tasks remain the same for the duration of an experiment if it should run for several days. Tasks may not be traded because each one teaches specific skills that all students need to learn.

The **recorder** begins the experiment by reading the entire set of directions to the other members while they follow along on their sheets. When they have finished reading, the recorder begins to prepare the lab sheet that the group will turn in for a grade. The lab sheet must include the job performed by each member (5 points), the purpose of the experiment (5 points), the questions (1 point each), and the answers (2 points each).

The other three members of the group are responsible for supplying the answers to the questions as the recorder writes them down.

The **controller** follows the procedure outlined in the directions, maintains discipline within the group, guides the others through the activities, and practices being a good leader without being the boss. The controller also supplies answers for the lab sheet.

The **hopper** wears a colored cord around her neck to symbolize her job because she is the only member of the group allowed out of her seat. The hopper is responsible for all materials. She will ensure that the team receives all the materials listed in the text at the beginning of the experiment, and that the team returns all the materials at the end of the experiment. She washes and dries all tools and puts them away. She notifies the teacher if anything is missing or broken. The hopper also supplies answers for the lab sheet.

The **observer** supports the other members of the group by complimenting them for doing a good job. He substitutes for absent members and informs the group when it is time to clean up. The observer also supplies the answers for the lab sheet.

Five-Minute Clean-up

The five-minute clean-up is a time of hustle, bustle, and noise, but everyone has a specific set of tasks that must be completed.

The **recorder** turns in the lab folder containing the completed lab sheet. The **controller** clears the work area, washing and drying the table top as necessary. The **hopper** accounts for all materials and equipment, washes and dries them as necessary, and returns them to their proper locations. The **observer** straightens chairs and helps the hopper put away materials and equipment.

Lab 1　OBSERVING INDIRECTLY

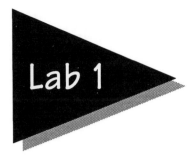

Recorder _____

Controller _____

Hopper _____

Observer _____

Purpose

To study atoms indirectly by observing the effects of soap particles on a light beam

Materials

2 baby food jars
piece of soap
water
flashlight
spoon

Procedure

Step 1:　Nearly fill the two jars with water. Add a piece of soap about 1/4 the size of a pea to one jar. Stir the mixture until the soap dissolves. Do nothing to the other jar. Answer the following questions.

　　　1.　Which jar is the experimental jar? _____

　　　2.　Why is a control jar necessary? _____

　　　_____

　　　_____

Step 2: All groups must do this step simultaneously, so raise your hands when your group is ready to turn off the lights. In the darkened room, direct the flashlight beam against the side of the jar with the soapy water. Observe what happens to the light beam as it passes through the soapy water. Then respond to the following:

3. Scientifically describe what you see in the jar of soapy water. List as many observations as you can.

4. Does the light beam go straight through the jar of soapy water or does it come out in all directions?

Step 3: Now direct the flashlight beam into the jar of clear water. Observe the effects of the clear water on the light beam and respond to the following:

5. Scientifically describe what you see in the jar of clear water.

6. Does the light beam go straight through the jar of clear water or does it come out in all directions?

7. In which jar of water does the light beam shine the greatest distance? Why?

8. Which jar of water reflects the most light? Why?

9. What are you observing indirectly in this experiment?

Team Science © 1996 Zephyr Press, Tucson, AZ

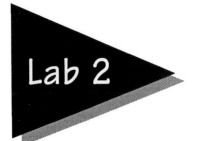

Lab 2   SHOEBOX ATOMS

Recorder   _____

Controller   _____

Hopper   _____

Observer   _____

Purpose

To observe indirectly the properties of shoebox atoms and to identify their contents

Materials

At least one sealed shoebox with one or more objects from home for each lab group

Procedure

Step 1: Label each shoebox with your lab number and a letter.

Step 2: Carefully tilt, rattle, turn over, and shake the shoebox atom. Discuss what you hear with the other members of your team. Then, using the data sheet, record your answers to the following questions:

1. How many objects are in the shoebox atom?

2. How big is each item?

3. What shape (cube, sphere, pyramid, irregular, etc.) is each item?

4. Is each item hard, soft, or both?

5. How many flat sides does each item have?

6. Of what material (plastic, rubber, wood, glass, metal, etc.) is each item made?

7. How does each item move (rolls smoothly, rolls roughly, slides, a combination of several movements) when the shoebox is tilted in various directions?

8. Draw the shape of each item.

Step 3: Repeat this procedure with as many shoebox atoms as time allows. Then respond to the following:

9. How is this lab similar to what scientists do when they study real atoms?

10. How is this lab different from what scientists do when they study real atoms?

Shoebox Atom Data

	Box A	Box B	Box C
No. of Items			
Size			
Shape			
Hardness			
No. of Flat Sides			
Material			
Movement			
Drawing			

Team Science © 1996 Zephyr Press, Tucson, AZ

Lab 3

PHYSICAL PROPERTIES

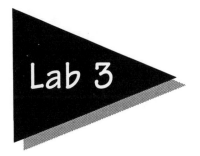

Recorder _____

Controller _____

Hopper _____

Observer _____

Purpose

To identify and classify elements on the basis of their physical properties

Materials

safety glasses
electric circuit
alcohol burner
forceps
sandpaper

Elements

aluminum
copper
sulfur
carbon
iron
iodine
magnesium

Procedure

Record the following information on the data sheet.

Part 1: Physical Properties

Step 1: Write the symbol for each element in the appropriate space. Refer to the Periodic Table if necessary.

Step 2: Study the appearance of each element. When possible, sand the sample with sandpaper to remove any tarnish. Record the color, shape, luster, and texture.

Step 3: Try to bend each element. If it bends without breaking, it is malleable. Write "Yes" or "No" in the malleability column.

Step 4: Use the electric circuit to test the elements for electrical conductivity. Sand away any tarnish for a clean contact surface. Touch each element with the two wires of the circuit. Write "Yes" if the light goes on and "No" if it doesn't.

Part 2: Heating Behavior

Step 1: Because of the toxic properties of sulfur and iodine, the teacher will demonstrate the heating behavior of these two elements. Observe carefully, and record the data in the last column under "1. Observations while heating" and "2. Observations after cooling."

Step 2: Using forceps, hold each of the remaining elements directly in the flame for 2 minutes.

> ## DO NOT LOOK DIRECTLY AT THE MAGNESIUM WHILE IT IS BURNING!

While the element cools, record your observations of the heating process in space 1. After the element has cooled, record your observations in space 2. Then respond to the following:

- Which element requires the most heating to show any visible change?

- Which element requires the least heating to change?

Physical Properties Data Sheet

	Symbols	Description	Malleability	Conductivity	1. Observations while heating.		2. Observations after cooling.	
					1.	2.	1.	2.
Sulfur								
Iodine								
Carbon								
Copper								
Iron								
Aluminum								
Magnesium								

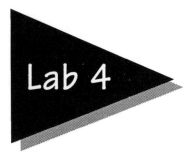

Lab 4   POP-TOP ROCKETS

Recorder _____

Controller _____

Hopper _____

Observer _____

Purpose

To study and compare the energy produced by sodium bicarbonate and water

Materials

1 film container
2 sodium bicarbonate tablets
water
meter stick
graduated cylinder
safety glasses
medicine dropper

Procedure

Step 1:   Remove the two sodium bicarbonate tablets from the packet. Carefully break the tablets into five pieces (three-halves and two-quarters) as follows: Break tablet one in half. Then break one-half in half again. Break tablet two in half. Give one-half to the teacher.

Step 2:   Pour 15 mL of water in the film container and place the container on your designated launch site outside.

Team Science © 1996 Zephyr Press, Tucson, AZ

Step 3: Before continuing, all team members must put on their safety glasses.

Step 4: Drop one-quarter piece of sodium bicarbonate into the film container, immediately snap the cap into place and stand back.

Step 5: Measure the distance from the launch site to where the cap first touched down and record it on your data sheet.

Step 6: Wash and dry the container between each test as you repeat steps 2 through 5, using one-half tablet for test 2 and three-quarters tablet for test 3. Record the distances on the data sheet. Remember to wear your safety glasses.

Step 7: Respond to the following:

1. What caused the cap to pop off the container?

2. What kind of gas is formed when H_2O combines with NaC_2?

3. Is this an example of a chemical change or a physical change?

4. How much sodium bicarbonate would be needed to launch the cap seven meters?

Pop-Top Rocket Data

Test	Sodium bicarbonate	Distance
1	$1/4$ tablet	
2	$1/2$ tablet	
3	$3/4$ tablet	

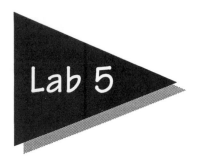

Lab 5

MYSTERY POWDERS

Recorder _____

Controller _____

Hopper _____

Observer _____

Purpose

To identify unknown substances by examining their physical and chemical properties when exposed to several common reagents

Materials

10 unknown powders
6 paper plates
toothpicks
baby food jar
paper towel
ruler
medicine dropper

Reagents

water
vinegar
iodine

Procedure

Step 1: Use the ruler to divide each paper plate into 5 wedges. Number the wedges from 1 to 10.

Step 2: Fill the baby food jar with water to wash your toothpick in after each test.

Step 3: Place one scoop of powder in each wedge on your plates. Be careful to match the numbers on the jars with the numbers on your wedges.

Team Science © 1996 Zephyr Press, Tucson, AZ

Step 4: Mound each powder into a tiny volcano with a crater.

Step 5: Place one drop of water into the crater in each mound of powder. Stir each powder with a clean, dry toothpick. (Wash the toothpick in the baby food jar and dry it with a paper towel after each application.) Any powder left on the toothpick could contaminate your results.

Step 6: Record the results of your tests on the data sheet and respond to the following:

    1. Which powders were insoluble in water?

Step 7: Dispose of the paper plates and repeat steps 1 through 6 using vinegar as the reagent. Respond to the following:

    2. Which powders bubbled?

Step 8: Dispose of the paper plates and repeat steps 1 through 6 using iodine as the reagent. Respond to the following:

    3. Which powders turned black?

    4. After your teacher has revealed the names of the mystery powders, write the names in a list. Next to each name describe how it reacts with each reagent.

    5. Which powder do you think would be the easiest to identify? Why?

    6. Which powder do you think would be the most difficult to identify? Why?

Mystery Powders Data Sheet

Powders	Water	Vinegar	Iodine
1			
2			
3			
4			
5			
6			
7			
8			
9			
10			

TEACHER'S INSTRUCTIONS

Before the lab begins, label the tops and bottoms of 10 jars before filling them with the following:

1. granulated sugar
2. cornstarch
3. baking powder
4. salt
5. chalk dust
6. baking soda
7. talcum powder
8. white flour (wheat)
9. powdered sugar
10. powdered antacid tablets

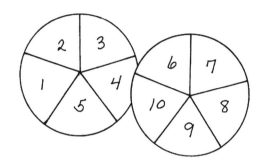

Use ice cream sticks as scoops to measure the amount of powder to be used for each test. Label the ice cream sticks with matching numbers so that as students help themselves to the powders there will be less possibility of contamination.

Reveal the identity of the powders after the tests have been completed.

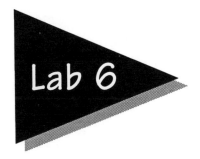

Lab 6

MIXTURE VS. COMPOUND

Recorder _____

Controller _____

Hopper _____

Observer _____

WEAR GLASSES!

Purpose

To learn how to distinguish between a mixture and a compound

Materials

safety glasses

magnet

iron filings

alcohol burner

test tube tongs

scoop

3 test tubes

paper towels

magnifying lens

sulfur powder

2 sheets smooth paper

Procedure

Step 1: Use the magnifying lens to study the appearance of the iron filings and sulfur powder. Then respond to the following:

1. List the physical properties (color, odor, texture, etc.) of the sulfur powder.

2. List the physical properties of the iron filings.

Step 2: Test the solubility of the iron filings and sulfur powder in water. Halfway fill two test tubes with water. Add a small amount of iron filings to one test tube and a small amount of sulfur powder to the other. Swirl each test tube for one minute. Respond to the following:

3. Describe what you observed in the test tube with the iron filings.

4. Describe what you observed in the test tube with the sulfur powder.

5. Could water be used to separate iron filings and sulfur powder in a mixture? Explain.

Step 3: Dispose of the contents of the test tubes by dumping them into the container on the sink. Wash and dry the test tubes before continuing with the experiment.

Step 4: Place a small amount of iron filings on a sheet of clean, smooth paper. **DO NOT PUT THE MAGNET INTO THE FILINGS!** Move the magnet back and forth *under* the paper. Similarly test the sulfur powder with the magnet on another piece of clean, smooth paper. Respond to the following:

6. Describe how the iron filings react to the magnet.

7. Describe how the sulfur powder reacts to the magnet.

8. Could a magnet be used to separate a mixture of iron filings and sulfur powder? Explain.

Step 5: Mix a small amount of sulfur powder and iron filings on a paper towel. Try to separate the iron filings and sulfur powder mixture by holding the magnet under the towel. Respond to the following:

9. Does the magnet separate the iron filings from the sulfur powder? Explain.

Team Science © 1996 Zephyr Press, Tucson, AZ

Step 6: Pour the iron filings and sulfur powder mixture into a clean, dry test tube.

> POINT THE TEST TUBE AWAY FROM PEOPLE
> IN THE NEXT PART OF THIS EXPERIMENT;
> THE FUMES ARE POISONOUS.

Use tongs to hold the test tube while heating the mixture until it glows red, then heat it 2 more minutes.

Step 7: After the test tube has cooled, remove the contents by gently tapping the test tube upside down on a clean, smooth paper. Respond to the following:

10. Describe the material removed from the test tube.

Step 8: Test the substance with a magnet by holding the magnet under the paper. Respond to the following:

11. Is the material attracted to the magnet?

12. Can the magnet be used to separate the iron filings from the sulfur powder?

13. Describe the odor of the material.

14. Is the material a mixture or a compound? Explain.

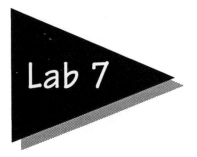

Lab 7

SEPARATING A MIXTURE

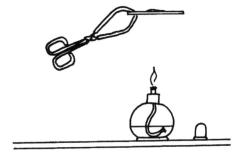

Recorder _____

Controller _____

Hopper _____

Observer _____

Purpose

To identify physical changes; to learn how to reclaim materials from a mixture by evaporation, distillation, and filtration

Materials

safety glasses
sand
alcohol burner
ice cream stick scoop
baby food jar
sugar
graduated cylinder
flask
water
rubber stopper and tubing

medicine dropper
test tube
glass slide
filter paper
funnel
ring stand and iron ring
tongs
wire screen
magnifying lens

Procedure

Part 1

Evaporation is used to separate a material that is dissolved in a liquid from the liquid.

Team Science © 1996 Zephyr Press, Tucson, AZ

Step 1:   Use the ice cream stick to measure 3 scoops of sugar into the baby food jar. Closely examine the sugar with the magnifying lens. Respond to the following:

   1.   What physical properties of sugar can you see with a magnifying lens?

Step 2:   Add 25 mL of water to the sugar in the jar and swirl the contents. Respond to the following:

   2.   What happened to the sugar when it was mixed with the water?

> **ALL TEAM MEMBERS MUST WEAR SAFETY GOGGLES.**

Step 3:   Use the medicine dropper to place 2 drops of the sugar water on a clean glass slide.Slowly dry the slide by holding it *high* above the heat of the alcohol burner. (The slide will break if it gets too hot.) Respond to the following:

   3.   Describe the material on the glass slide after the liquid has evaporated.

   4.   Taste the material. Does it still taste sweet like sugar?

   5.   How does this experiment show that dissolving sugar in water is a physical change?

Part 2

Distillation is used to separate materials with different boiling points.

Step 1:   Set up the distillation apparatus as shown.

Step 2:   Dissolve 3 scoops of sugar in 25 mL of water. Heat the sugar water in the flask until it boils and collects in the dry test tube. Continue distilling the solution until you have enough to taste.

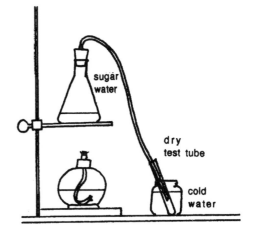

Remove the tubing from the test tube before extinguishing the alcohol burner or the material collected in the test tube will return to the flask.

Step 3: Taste the liquids in the flask and test tube by dipping in the end of your little finger. Respond to the following:

6. What does the liquid in the flask taste like?

7. Is the liquid in the test tube sweet? Which of the original materials is in the test tube?

8. Based on the results of this experiment, which of the original materials has the lowest boiling point? (Which one boils first?)

9. Would distillation work if both materials had the same boiling point?

Part 3

Filtration is necessary to separate a material that won't dissolve from one that will.

Step 1: Set up the filtration apparatus as shown.

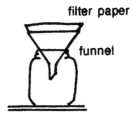

filter paper

funnel

Step 2: Mix 3 scoops of sugar with 2 scoops of sand in a clean, dry test tube. Fill the test tube half full of water. Swirl the contents. Respond to the following:

10. What happens to the sugar?
What happens to the sand?

Step 3: Place your thumb over the opening of the test tube and shake it upside down to get the sand out of the bottom of the test tube. Quickly pour the contents into the filter. When all liquid has passed through the filter, pour another 10 mL of water into the filter. Carefully open the filter paper and observe the contents. Respond to the following:

11. What remains in the filter paper? Why?

Step 4: Place 2 drops of filtrate (liquid that passed through the filter) onto a clean slide. Everyone must put on safety goggles before slowly drying the slide high above the heat so it doesn't break. Respond to the following:

12. What remained on the slide after drying?

13. Was the separation of the sugar-sand mixture a physical change? Explain.

14. What process could collect the water from the sugar water in the filtrate?

Lab 8

SUGAR CRYSTALS, PART 1

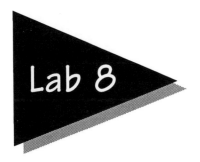

Recorder   _____

Controller   _____

Hopper   _____

Observer   _____

Purpose

To grow sugar crystals from a super saturated solution; to prove that sugar molecules remain sugar molecules even after they dissolve

Materials

tall baby food jar
string
scissors
paper clip
ice cream stick
sugar
water

Procedure

Step 1: Label the ice cream stick with your class period and group number.

Step 2: Tie one end of the string to the paper clip. Rest the stick across the rim of the jar and tie the other end of the string around it so that the paper clip doesn't touch the bottom of the jar.

Step 3: Take the prepared jar to the teacher to be filled with super saturated sugar solution. Sugar crystals will form on the surface of the solution, on the bottom of the jar, and on the rough surfaces of the string and paper clip. Respond to the following:

1. Where do you think the crystals will form first? Why?

2. What is the purpose of the paper clip?

3. What do you think will happen if the paper clip is too close to the bottom of the jar?

4. What 2 compounds are in the solution and what are their formulas? (Use reference material if necessary.)

5. Is the sugar solution a mixture or a compound? Explain.

6. If there are 500 mL of sugar for every 250 mL of water, what is the ratio of sugar to water?

Step 4:   Periodically observe the progress of your crystals for the next 6 to 8 weeks, but try not to disturb any suspended crystals. Answer the following questions as you learn the answers:

7. How much time passed before you observed crystals forming?

8. Where did the crystals form first? Was the hypothesis you set forth in your answer to question 1, correct?

9. Are sugar molecules still present in the solution? Explain.

TEACHER'S INSTRUCTIONS

Before the lab begins, do the following:

- Mix 250 mL (1 cup) water with 500 mL (2 cups) sugar.

- Slowly bring to a rolling boil.

- Boil uncovered for 5 minutes.

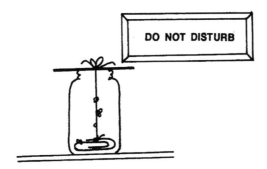

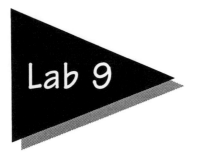

Lab 9

SUGAR CRYSTALS, PART 2

Recorder _____

Controller _____

Hopper _____

Observer _____

Purpose

To study and compare the geometric shape of several sugar crystals; to compare sugar crystals to salt crystals

Materials

sugar crystals
pocket microscopes
scissors
salt
black paper
hand lens

Procedure

Step 1: Examine your jar of sugar crystals. Respond to the following as you notice where the crystals grew, the patterns they formed, the various sizes, any changes in the original solution, and so on.

1. Where did most of the crystals in your jar grow?

2. Where were the largest crystals and the smallest crystals?

3. How has the solution changed (amount, viscosity, etc.)?

Step 2:   Pour the super saturated sugar solution in the sink. Remove the sugar crystals attached to the string and quickly rinse them. Drain the crystals on a paper towel and take them back to your group. Leave the baby food jar soaking in the sink. Respond to the following:

   4.   What will happen to the crystals left in the jar soaking in the sink?

Step 3:   Carefully break off 3 of the largest and most perfectly formed crystals to use in today's lesson. Place them on a piece of paper while you study them with the hand lens. Respond to the following:

   5.   How many sides does each crystal have?

   6.   Discuss how the 3 crystals are similar and how they are different.

Step 4:   Now place a perfect little crystal on the black paper so you can study it with the pocket microscope.

   7.   Sketch what you see through the pocket microscope.

Step 5:   Sprinkle about 10 grains of salt onto the black paper. Examine them with the pocket microscope. Respond to the following:

   8.   Discuss how the sugar and salt crystals are similar and how they are different.

   9.   What is the geometric shape of a salt crystal?

   10.   Predict what the crystals in a super saturated sugar-salt solution would look like.

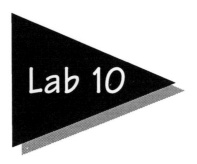

Lab 10

MINERAL IDENTIFICATION

Recorder _____

Controller _____

Hopper _____

Observer _____

Purpose

To become familiar with the properties of minerals and the tests used to identify them

Materials

Mohs Scale of Hardness
box of 12 minerals
magnet
graduated cylinder
medicine dropper
balance scale
hydrochloric acid
hand lens
baby food jar
staple

Procedure

Step 1: Record your observations on the data sheet as you test each mineral. The tests may be completed in any order, but run the same test on all of the minerals before proceeding to the next test.

Team Science © 1996 Zephyr Press, Tucson, AZ

Step 2: Record the exact color(s) of each mineral

Step 3: Under "luster," write "transparent" if you can see objects through the material, "translucent" if you can see only light through it, and "opaque" if you cannot see through it at all.

Step 4: Wash and dry the porcelain tile before beginning this test. Apply slightly more pressure than you would use when writing. Record the color of the streak. Write "colorless" if there is no streak.

Step 5: It is not necessary to break the mineral samples to do this test. Write "cleavage" if the sample has at least one flat surface. Write "fracture if all surfaces are rough and irregular.

Step 6: Be sure the minerals in the Mohs kit are in order. (Ask the teacher to check them if you aren't sure.) Strong pressure must be applied to scratch the minerals. Always begin by trying to scratch 5 (Apatite) with the mineral from your box. Proceed up the scale to 6 if the mineral scratches 5, and down the scale to 4 if 5 scratches your mineral. Continue testing the minerals in the Mohs kit with your mineral until you find one that cannot be scratched and cannot scratch your mineral. Record the number of the mineral from the Mohs kit on your data sheet.

Step 7: Weigh the mineral on the balance scale (left side) to find the mass. Use the medicine dropper to measure exactly 20 mL of water in the graduated cylinder and slide in the mineral. The volume is the amount that the water rises. Dividing that number into the mass will give you the specific gravity in g/cm^3.

$$\text{Specific gravity} = \frac{\text{Mass}}{\text{Volume}}$$

Step 8: Place the minerals on a paper towel to do this test. Drop *one* drop of acid on each mineral. **DO NOT GET ACID ON ANYTHING BUT THE MINERALS!** Write "carbonate" if the mineral bubbles and write nothing if it doesn't.

Step 9: Note which mineral samples are attracted to a magnet. Test to see whether any samples are magnetic by attempting to pick up a staple. Write this information on the back of the data sheet.

Step 10: Match the data on your data sheet with the data for the minerals on the mineral identification table.

Mineral Identification Data Sheet

	Color	Luster	Streak	Breakage Pattern	Hardness	Specific Gravity	Acid Test	Mineral
a.								
b.								
c.								
d.								
e.								
f.								
g.								
h.								
i.								
j.								
k.								
l.								

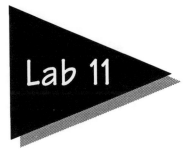

Lab 11

WATER AS A SOLVENT

Recorder _____

Controller _____

Hopper _____

Observer _____

Purpose

To compare the solvency of water with that of other solvents

Materials

medicine dropper
ice cream stick scoop
drinking straw
graduated cylinder
3 test tubes
baby food jar

Solvents

water
vinegar
oil

Solutes

cornstarch
baking soda
sugar
powdered milk

Procedure

Step 1: Pour 20 mL of each solvent into a test tube and label it accordingly. The oil can be poured directly into the test tube so that cleaning the graduated cylinder will be easier. Hold the test tube of water next to the test tube marked oil as you pour the oil to the same height. Use the medicine dropper for a perfect match. Respond to the following:

    1. Which solvent do you think will dissolve the solutes most easily?

    2. Which solvent do you think will dissolve the fewest solutes?

Step 2: Add 1 scoop of cornstarch to each test tube.

Step 3: Before stirring, record your observations in the appropriate spaces on the data sheet.

Step 4: After stirring the contents of each test tube with a clean straw for 10 seconds, record your observations in the appropriate spaces on the data sheet.

Step 5: Pour the oil into the container on the sink, and pour the vinegar and water down the drain. Thoroughly wash and rinse the test tubes before continuing with the next test.

Step 6: Repeat steps 1 through 5 using baking soda.

Step 7: Repeat steps 1 through 5 using sugar.

Step 8: Repeat steps 1 through 5 using powdered milk. Respond to the following:

    3. In which solvent did the solutes dissolve most readily?

    4. In which solvent did the solutes dissolve with the greatest difficulty?

Water as a Solvent Data

	Water		Vinegar		Oil	
Cornstarch	Before	After	Before	After	Before	After
Baking Soda	Before	After	Before	After	Before	After
Sugar	Before	After	Before	After	Before	After
Powdered Milk	Before	After	Before	After	Before	After

Lab 12

SALINITY AND TEMPERATURE

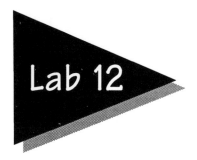

Recorder   _____

Controller   _____

Hopper   _____

Observer   _____

Purpose

To summarize the relationship between salinity and water temperature

Materials

graduated cylinder
spoon
balance scale
salt
2 sheets of paper, each 6 square cm
hot water
thermometer
ice water
2 baby food jars

Procedure

Step 1: Place one sheet of paper on the left side of the balance scale. Set the slider at 0 and balance the scale using the adjustment knob. This procedure will ensure that you don't include the weight of the paper in the weight of the salt.

Team Science © 1996 Zephyr Press, Tucson, AZ

Step 2: Set the slider at 1 gram and add salt to the paper until the scale balances. Respond to the following:

    1. On which side of the balance scale is the salt?

    2. Why is it desirable to adjust the balance of the scale with just the paper on it?

Step 3: Pour 100 mL of ice water into the baby food jar and measure its temperature. Record the temperature on the data chart. Label your answer with °C.

Step 4: Add salt to the water, 1 g at a time, until no more salt will dissolve. Stir the water thoroughly after each addition and allow time for excess salt to settle on the bottom. Record on the data chart the number of grams of salt dissolved in the ice water.

Step 5: Fill the other baby food jar with 100 mL of hot water and measure its temperature. Record the temperature of the hot water. Label the answer °C. Respond to the following:

    3. Do you think that more salt or less salt will dissolve in the hot water?

Step 6: Now repeat the procedure outlined in step 4 using the hot water. Continue to enter all data on the data chart. Respond to the following:

    4. Which temperature of water dissolved the most salt? Was your hypothesis correct?

Salinity vs. Temperature Data

	Temperature	Amount of Salt
Hot Water		
Cold Water		

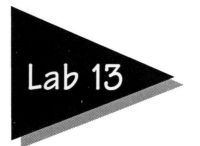

Lab 13 GASTROPOD RACE

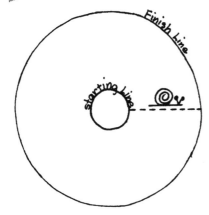

Recorder _____

Controller _____

Hopper _____

Observer _____

Purpose

To observe the physical characteristics of gastropods; to study the effect of various stimuli on their behavior

Materials

4 garden gastropods
race track pattern
masking tape
paper, 31 cm by 31 cm
clock with second hand
containers for keeping gastropods safe for next day

> **NOTE:** Containers need secure lids, abundant air holes, moist vegetation, few gastropods, and identification. Keep the containers in a cool location (65–70°F) for best results. You will be responsible for lost, ill, or deceased contestants.

Procedure

Step 1: Trace the race track pattern onto a piece of paper. Label the circle in the center Starting Line and the circumference Finish Line.

Step 2: Name each gastropod and attach a tiny piece of tape with its initials to its shell.

Team Science © 1996 Zephyr Press, Tucson, AZ

Step 3: Discuss with your group the possible effects on the gastropods of various stimuli such as light, dark, noise, quiet, lettuce, cucumber, moisture. List the stimuli. You will test each stimulus on your list during the practice races. Respond to the following:

1. List the stimuli that your group thinks would best motivate your gastropods to move in the desired direction.

Step 4: Select a time keeper who will be responsible for starting the race and who will tell the recorder the starting time so the recorder can record it on the data sheet.

Step 5: Place the gastropods inside the starting circle. The time keeper should begin the race when all contestants and their trainers are ready. Test the various stimuli discussed earlier, but do not touch the gastropods except to separate them if one crawls on top of another. Each trainer is responsible for telling the recorder the exact time her gastropod reached the Finish Line so that the time may be entered on the data sheet.

Step 6: Subtract the beginning time from the finish time and record the result on the data sheet. Respond to the following:

2. Which stimulus worked best to move the gastropod in the desired direction? How did the result compare to your hypothesis in step 3?

3. Which stimulus made the gastropods stop or move in another direction?

4. What was the fastest time for your practice race?

Step 7: Enter your speediest gastropod in the final race at the end of the period. Good luck to all.

Step 8: Return all gastropods to their containers until tomorrow.

Gastropod Race

Snail's Name	Beginning Time	Finish Time	Total Time

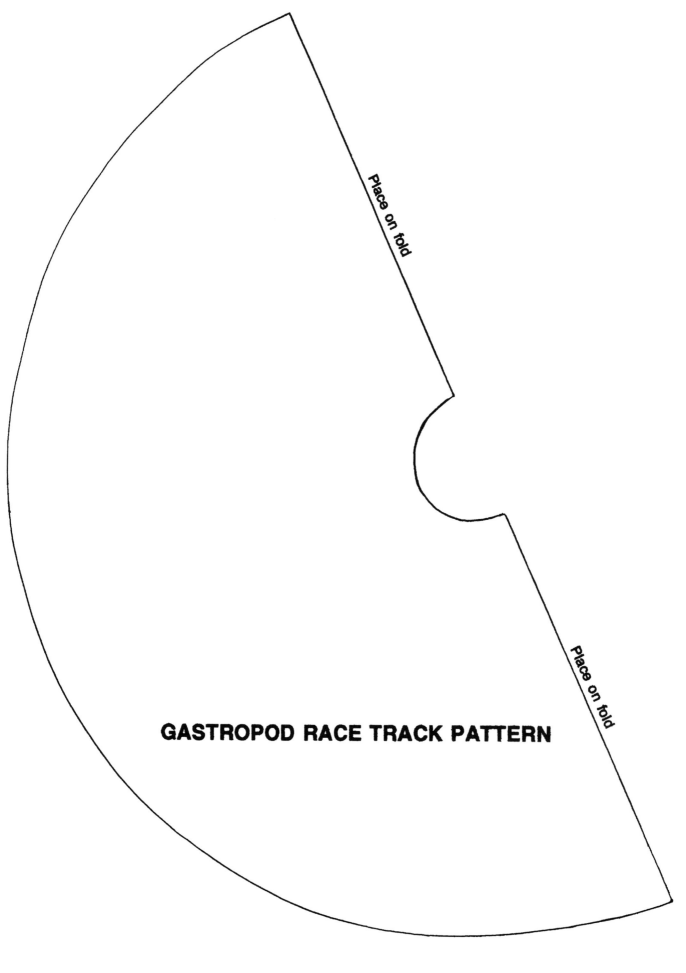

Place on fold

Place on fold

GASTROPOD RACE TRACK PATTERN

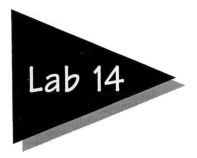

Lab 14

GASTROPOD CLIMB

Recorder _____

Controller _____

Hopper _____

Observer _____

Purpose

To observe the climbing ability of gastropods; to study the effect of various stimuli on their climbing ability

Materials

4 gastropods
hinged chip board
masking tape
clock with second hand
containers for keeping gastropods safe for next day

NOTE: Containers need secure lids, abundant air holes, moist vegetation, few gastropods, and identification. You will be responsible for lost, ill, or deceased contestants.

Procedure

Step 1: Stand the chip board between two books so that the walls are as vertical as possible.

Step 2: Name each gastropod and attach a tiny piece of tape with its initials to its shell.

Step 3: Discuss with your group the possible effects on the gastropods of various stimuli such as light, dark, noise, quiet, lettuce, cucumber, moisture. List the stimuli. You will test each stimulus on your list during the practice races. Respond to the following:

    1. List the stimuli that your group thinks would motivate your gastropods to move in the desired direction.

Step 4: Select a time keeper who will be responsible for starting the race and who will tell the recorder the starting time so she can record it on the data sheet.

Step 5: Place the gastropods at the bottom of the climbing wall so that their shells touch the table. The time keeper should begin the race when all contestants and their trainers are ready. Test the various stimuli discussed earlier, but do not touch the gastropods except to separate them if one crawls on top of another. Each trainer is responsible for telling the recorder the exact time his gastropod reached the top of the climbing wall.

Step 6: Subtract the beginning time from the finish time to calculate the total time and record it on the data sheet. Respond to the following:

    2. Which stimulus worked best to move the gastropod in the desired direction? How did the results compare to your hypothesis in step 3?

    3. Which stimuli made the gastropods stop or move in another direction?

    4. What was the fastest time for the practice climb?

Step 7: Now place your gastropods at the top of the wall and race them to the bottom. Respond to the following:

    5. Record here what you observed as the gastropods raced to the bottom of the wall.

Step 8: Enter your speediest gastropod in the final climb at the end of the period.

Step 9: Return all gastropods to their containers until tomorrow.

Team Science © 1996 Zephyr Press, Tucson, AZ

Gastropod Climb

Snail's Name	Beginning Time	Finish Time	Total Time

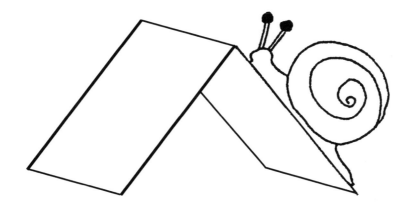

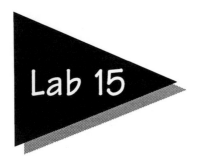

Lab 15

GASTROPOD TOWING SERVICE

Recorder   _____

Controller   _____

Hopper   _____

Observer   _____

Purpose

To observe the strength of a gastropod; to calculate the relative strength of a person

Materials

1 large gastropod
baby food jar
balance scale and weights
4 paper clips
1 rubber band
1 sheet of paper, 6 square cm
heavy objects:

- rocks
- nuts and bolts
- scissors
- marbles
- other assorted heavy items

Team Science © 1996 Zephyr Press, Tucson, AZ

Procedure

Step 1: Place the paper on the left side of the balance scale. Set the slider at 0 and balance the scale using the adjustment knob. This procedure will ensure that you don't include the weight of the paper in the weight of the gastropod. Record the weight of the gastropod on the data chart.

Step 2: String the 4 paper clips and the rubber band together to form a chain. Tape the rubber band to the gastropod's shell and the paper clips to the baby food jar.

Step 3: Gradually fill the baby food jar with objects until the gastropod can no longer move the jar. Remove the object that made the jar too heavy to move before weighing the jar and its contents.

Step 4: Carefully remove the rubber band from the gastropod's shell. Weigh the tape, rubber band, paper clips, baby food jar and all of its contents. Record the total weight on the data sheet.

Step 5: To calculate how many times its own weight a gastropod can tow, divide the gastropod's weight into the total towed weight. Round the answer to the nearest 10th. Respond to the following:

1. If a person weighed 100 pounds, how much weight would he have to tow to be as strong as a gastropod?

Step 6: Wash and dry containers and provide fresh lettuce. It is your responsibility to provide a safe and healthful environment for your gastropods. Return all gastropods to their containers until tomorrow.

Gastropod Towing Service

Snail's Name	Snail's Wt.	Maximum Towed Wt.	Towed Wt. / Snail Wt.

Lab 16 GASTROPOD FEAST

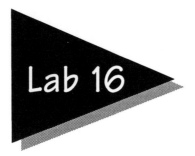

Recorder _____

Controller _____

Hopper _____

Observer _____

Purpose

To observe a gastropod's eating habits

Materials

4 garden gastropods
balance scale
lettuce
clock with second hand
1 sheet of paper, 6 square cm

Procedure

Step 1: Record each gastropod's name on the data sheet.

Step 2: Place the sheet of paper on the left side of the balance scale. Set the slider at 0 and balance the scale using the adjustment knob. This procedure will ensure that you don't include the weight of the paper in the weight of the lettuce. Record the weight of each piece of lettuce next to the name of the gastropod that will eat it.

Team Science © 1996 Zephyr Press, Tucson, AZ

Step 3: Place each gastropod on a piece of lettuce and, as soon as the snail begins eating, time 15 minutes. Each trainer must monitor the time.

Step 4: Turn the leaf over so that you can closely observe the gastropod eating. Respond to the following:

1. Where is the gastropod's mouth located?

2. What does the inside of the gastropod's mouth look like?

Step 5: After the gastropod has eaten for 15 minutes, weigh the leaf again and record the weight on the data sheet.

Step 6: To calculate how much of the leaf the gastropod has eaten, subtract the final leaf weight from the beginning leaf weight. Record the difference on the data sheet.

Step 7: Your gastropods have successfully completed their assignments and have earned their freedom. Release your gastropods in the designated area or take them home after school. You are responsible for their safe return to nature. Wash and dry all containers.

Gastropod Feast

Snail's Name	Leaf Wt Before	Leaf Wt. After	Amount Eaten

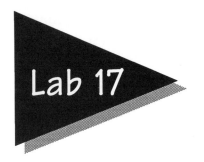

Lab 17

INTRODUCTION TO THE MICROSCOPE

Recorder   _____

Controller   _____

Hopper   _____

Observer   _____

Purpose

To learn the parts of a microscope and how to operate the microscope; to learn how to prepare a wet mount slide

Materials

microscope
tweezers
glass slide
scissors
cover slip
piece of newspaper
water
medicine dropper
hand lens

Procedure

Step 1: Cut out a lower case *e* from the newspaper. Use the tweezers to place it on the glass slide. Prepare a wet mount slide by placing a drop of water on the *e* and covering it with a cover slip.

Respond to the following:

1. What happened to the drop of water when the cover slip was placed on the *e?*

2. What is the function of the water in preparing the slide?

Step 2: Examine the *e* with the hand lens.

3. Draw the *e* as it appears through the hand lens.

Step 3: Place the *e* right-side up over the hole in the stage of the microscope. Hold it in place with the clips.

Step 4: Set the diaphragm on the largest opening possible.

Step 5: Click the low power lens into viewing position (directly over the hole in the stage). Use the coarse adjustment knob to lower the lens as close to the slide as possible without touching it.

Step 6: While looking through the eyepiece adjust the mirror so that you can see light. Still looking through the eyepiece, slowly raise the lens until the *e* is in focus. (If you can't see the letter, move the slide to be sure the letter is under the lens.) Finally, adjust the fine focus until the *e* is clearly visible. Respond to the following:

4. In what direction is the *e* facing?

5. Draw the *e* as it appears under low power.

Step 7: Move the slide left, right, up, and down as you look through the microscope. Respond to the following:

6. In what direction did the *e* appear to move when you moved the slide to the left?

7. In what direction did the *e* appear to move when you moved the slide up?

Step 8: Move to a higher power lens, and repeat steps 5 and 6. Respond to the following:

8. Draw the *e* as it appears under high power.

9. Why was less of the letter visible under high power than under low power?

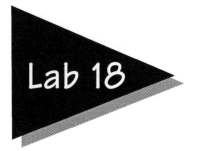

Lab 18   SUPER POWERS

Recorder   _____

Controller   _____

Hopper   _____

Observer   _____

Purpose

To become familiar with the magnification powers of a microscope

Materials

microscope
3 objectives

Procedure

Step 1:   Before studying the magnification powers of the microscope, use the labeled diagram to memorize all of its parts.

    1.   When you have memorized the names of the parts and can identify them for the other members of your group, write them in the correct places on the large microscope diagram.

Step 2:   Examine the eyepiece.

    2.   What is the number on the eyepiece?

Step 3:   Examine the microscope objectives.

    3.   What is the number of each objective?

      a. _____ , b._____ , c._____ .

Step 4: Find the total magnification of the microscope by multiplying the number on the eyepiece by the number on each of the objectives.

4. What is the power of magnification for each of the objectives?

Eyepiece		Objective		Power of Magnification
a._____	X	_____	=	_____
b. _____	X	_____	=	_____
c._____	X	_____	=	_____

Microscope Study Diagram

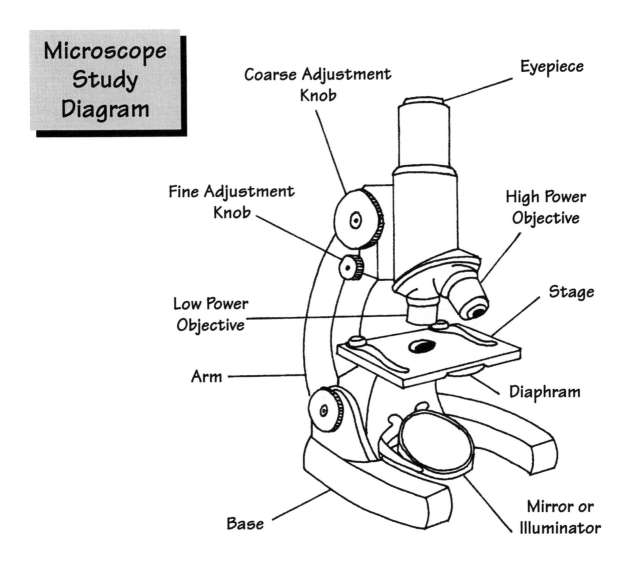

Coarse Adjustment Knob

Eyepiece

Fine Adjustment Knob

High Power Objective

Low Power Objective

Stage

Arm

Diaphram

Base

Mirror or Illuminator

Microscope Diagram

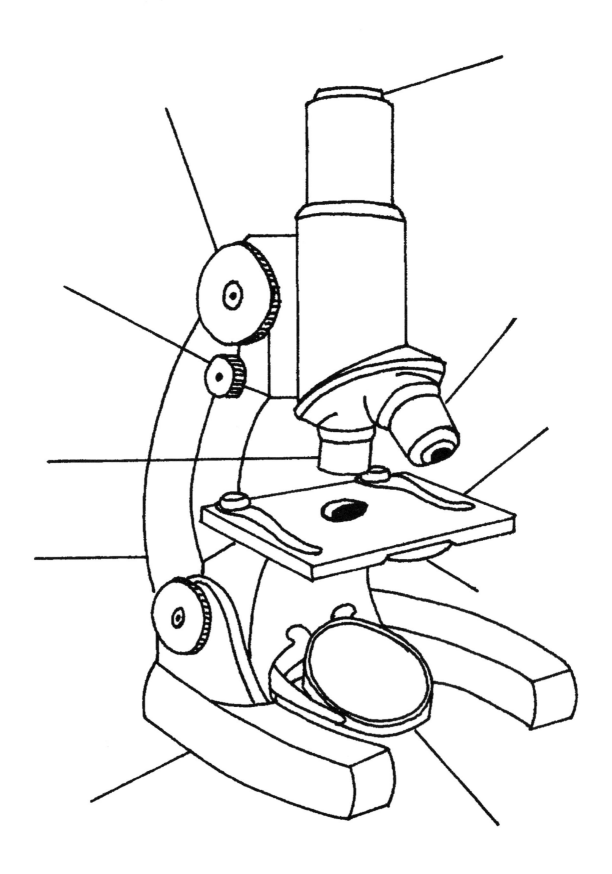

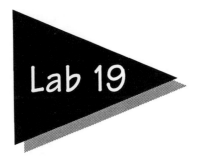

Lab 19 ▶ ONION CELLS

Recorder _____

Controller _____

Hopper _____

Observer _____

Purpose

To examine onion skin cells and to become familiar with the importance of staining tissue for microscopic study

Materials

compound microscope
medicine dropper
glass slide
onion
cover slip
methylene blue stain

Procedure

Step 1: Prepare a wet mount slide by removing the thin, transparent membrane from the inner surface of an onion scale. Smooth this membrane onto the glass slide and add a drop of water before covering it with a cover slip.

Step 2: Examine the slide with both low power and high power magnification. Respond to the following:

1. How many cells are visible through the microscope? (To facilitate counting, count the cells in 1/4 of the visible area and multiply by 4.)

    Low Power _____

    High Power _____

2. Draw a cell as it appears under high power.

3. With high power, how many lines are visible around the outside edge of each cell?

4. Can you see the nucleus with high power?

Step 3: Lift the cover slip and add a drop of methylene blue stain.

Step 4: Examine the slide under low power and high power magnification. Respond to the following:

5. Using high power, how many lines are visible around the outside edge of each cell?

6. How many dark, round structures inside the cell can you see with high power? Those are the nuclei.

7. How did the addition of the stain affect the visibility of the cell?

Team Science © 1996 Zephyr Press, Tucson, AZ

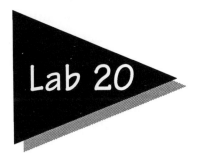

Lab 20 — CHEEK CELLS

Recorder   _____

Controller _____

Hopper   _____

Observer   _____

Purpose

To examine cheek cell structures; to compare cheek cell structures to onion skin cell structures

Materials

compound microscope
methylene blue stain
glass slide
medicine dropper
cover slip
flat-edged toothpick

Procedure

Step 1: Place a drop of methylene blue stain in the center of a glass slide.

Step 2: Collect cheek cells by gently scraping the inside of your cheek with a toothpick. (You won't be able to see the cells on the toothpick because they are microscopic.) Transfer the cells to the glass slide by stirring the drop of methylene blue stain with the toothpick containing the cells. Add a cover slip.

Step 3: Using your lowest power lens, move the glass slide around until you see a dark, irregularly shaped form. Center this dark "spot" before switching to highest power magnification.

Step 4: Adjust the gross focus by starting with the lens almost touching the glass slide. Begin backing off very slowly until you can see the cells. Finish focusing with the fine focus knob. Respond to the following:

1. How are these cells different from onion skin cells?

2. How many cells do you see?

3. How many lines do you see around the edge of each cell?

4. Can you see the nucleus (a dark round dot) inside the cells?

5. Draw one of the cells you see through the microscope. Label the parts you can see.

6. How is this cell different from the onion skin cells? How is it the same?

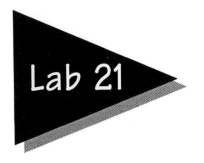

Lab 21

ENZYMES AND SODA CRACKERS

Recorder _____

Controller _____

Hopper _____

Observer _____

Purpose

To observe the chemical change that occurs when starch is combined with the enzymes in saliva; to recognize digestion as a chemical change

Materials

3 crackers (low salt)
2 test tubes
Benedict's solution
iodine stain
test tube tongs
graduated cylinder
medicine dropper
baby food jar
masking tape

Procedure

Step 1: Use masking tape to label one test tube "control" and the other "experimental."

Step 2: Test for the presence of starch by placing a drop of iodine stain on a soda cracker. If starch is present, the iodine stain will change color. Dispose of this cracker in the waste basket as soon as you have answered the next two questions.

  1. What color is the iodine on the soda cracker?

  2. Does the soda cracker contain starch?

Step 3: Fill the control test tube with a crushed soda cracker and stand it in the baby food jar.

Step 4: Fill the experimental test tube with a soda cracker that has been chewed for at least one minute and stand it in the baby food jar. (The more saliva mixed with the cracker the better.) Respond to the following:

  3. Why must the cracker in the experimental test tube be chewed?

  4. What is the function of the cracker in the control test tube?

Step 5: Add 5 mL of Benedict's solution to each test tube.

Step 6: Prepare a hot water bath by pouring 100 mL hot water into the baby food jar holding the 2 test tubes.

Step 7: When heated, sugar causes Benedict's solution to change color. Observe the color changes in the two test tubes for 10 minutes. Then respond to the following:

  5. Describe the color changes that occurred in each test tube.

  6. Which test tube exhibited the most color change? What caused more sugar to be present in that test tube?

  7. What tests for the presence of starch?

  8. What tests for the presence of sugar?

  9. What changed the starch in the soda cracker to sugar?

  10. Complete this statement:

  STARCH + _____ **=** _____ **+ WATER**

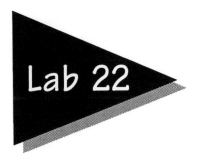

Lab 22 NON-NEWTONIAN GOOK

Recorder _____

Controller _____

Hopper _____

Observer _____

Purpose

To observe the effect of applied force on the viscosity of a non-Newtonian fluid

Materials

cornstarch
butter tub
water
2 cupcake liners
graduated cylinder
mixing spoon
tablespoon
balance scale
plastic bags

Procedure

Step 1: Place the 2 cupcake liners on the left side of the balance scale. Set the slider at 0 and balance the scale using the adjustment knob. This procedure will ensure that you don't include the weight of the liners in the weight of the cornstarch. Set the weight at 63 g and add cornstarch until the scale balances.

Step 2:   Pour 75 mL water into the butter tub. Slowly stir in 63 g of cornstarch. The mixture should be thick and shiny. (You may have to add a little more water or cornstarch to get the correct viscosity.) Respond to the following:

1.   Describe how the viscosity (thickness) of the mixture changed as you added the cornstarch.

Step 3:   Slowly sink your knuckles into the gook. "Grab" the fluid and pull up. Respond to the following.

2.   Describe what happened when you tried to pull up on the mixture.

Step 4:   Punch the gook with your fist. Respond to the following:

3.   Describe what happened when you punched the gook with your fist.

Step 5:   Roll the gook into a ball and hold it in the palm of your hand. Respond to the following:

4.   Were you able to roll the gook into a ball? What happened when you held it in the palm of your hand?

Step 6:   Slowly stretch it as far as you can.

Step 7:   Pull it apart quickly. Respond to the following:

5.   How does the speed at which the gook is pulled apart affect its behavior?

Step 8:   Stir the gook in a clockwise direction. Respond to the following:

6.   In which direction does the gook swirl after you stop stirring?

Step 9:   Test your own hypotheses by responding to the following:.

7.   List other physical properties that gook possesses.

Step 10: Clean all work surfaces. Wash and dry all tools and equipment. Put the gook in plastic baggies to take home or for disposal.

Team Science © 1996 Zephyr Press, Tucson, AZ

TEACHER'S INSTRUCTIONS

Introduce this experiment by reviewing the properties of fluids and Newton's law of viscosity.

A fluid is defined as anything that flows. It could be a gas, a liquid, an avalanche, or even a bunch of logs rolling together down a hill. All fluids possess a property known as "viscosity." Viscosity is what makes ketchup ooze rather than spray. It is the measurable thickness or resistance to flow in a liquid.

Newtonian fluids follow the law of viscosity first stated by English physicist Isaac Newton in the 1600s. The law states that the viscosity of a fluid can be changed only by altering the fluid's temperature. Honey is a common example. As the honey is warmed it flows more easily (becomes less viscous).

A non-Newtonian fluid defies Newton's law. These fluids have the same dependence on temperature, but their viscosity can also be changed by applying force or stress. In other words, if you slap, punch, or throw a non-Newtonian fluid against the wall, its viscosity increases tremendously. It won't splatter or ooze. For that instant it acts much like a solid—a piece of tough elastic.

Yank the gook and it will snap like a rubber band. Throw it on the floor and it will bounce.

Other non-Newtonian fluids can act in the opposite way. Force actually decreases their viscosity. Margarine, for example, is thick and waxy in a stick or tub. Yet, when spread over a piece of toast, it flows easily. The shearing force of the knife makes the margarine less viscous.

Use the Non-Newtonian Gook lab to reinforce the concepts you have taught. It may take several class periods to prepare a class for the lab, but the more familiar they are with Newton's law of viscosity, the more amazing the lab will appear. The day before the lab, you may want to demonstrate Newton's law of viscosity by pouring warm and cold honey.

Two boxes of cornstarch should be enough for eight lab groups, but it would be wise to have an extra box on hand in case someone measures incorrectly. You may find it helpful to go around adjusting each lab's gook with a little more water or cornstarch to get the correct consistency.

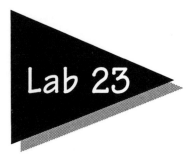

Lab 23   SOAP AND WATER

Recorder   _____

Controller   _____

Hopper   _____

Observer   _____

Purpose

To observe the effect of detergent on the surface tension of water

Materials

3 baby food jars
water
30 pennies or washers
clean medicine dropper
masking tape
paper towel
liquid dish detergent in dropper bottle

Procedure

Step 1: Label one jar "detergent" and a second jar "water." Use the
third jar to fill the first two jars almost to the rim with water.

Step 2: Add 3 drops of liquid detergent to the jar marked "detergent."

Step 3: Dry the rims of the jars with a paper towel before using the
clean, soap-free medicine dropper to finish filling the jars
exactly to the rim. Take care that the water does not rise above
the rim.

Team Science © 1996 Zephyr Press, Tucson, AZ

Step 4: Carefully add 1 penny to each jar filled with water without touching the water with your fingers. It is equally important that no one bumps the work table while you do this experiment or you will have to start over. Respond to the following:

1. Predict how many pennies you think each jar will hold before the water spills over the rim.

   Water: _____

   Detergent: _____

Step 5: Continue adding pennies to the jar marked "detergent" until the water spills over the rim. Respond to the following:

2. How many pennies fit into the jar with detergent in it?

Step 6: Continue adding pennies to the jar marked "water" until the water spills over the rim. Respond to the following:

3. How many pennies fit into the jar with just water in it?

4. Which jar held the most pennies? Why?

5. What allows the water to rise above the rim without spilling over?

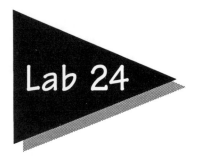

Lab 24

WATER ON A COIN

Recorder	_____
Controller	_____
Hopper	_____
Observer	_____

Purpose

To observe how strong surface tension is; to predict how many drops of water will fit on a coin

Materials

baby food jar
medicine dropper
1 penny
water
1 nickel
paper towels
sheets of colored paper, 6 square cm

Procedure

Step 1: Fill the baby food jar half full of water.

Step 2: Completely dry the penny, especially the edge, with a paper towel. Place the penny on a dry piece of colored paper.

Step 3: Use the medicine dropper to place 1 drop of water on the penny. Respond to the following:

Team Science © 1996 Zephyr Press, Tucson, AZ

1. Predict how many drops of water will fit on the penny before the water spills over the edge.

Step 4: Count the number of drops that you add before the water spills over the edge of the penny. Respond to the following:

2. How many drops fit on a penny? (Don't forget to count the first drop added in step 3.)

3. What holds so many drops of water together on top of the penny?

Step 5: Carefully dry your work surface. Now place a dry nickel on a dry piece of colored paper.

Step 6: Use the medicine dropper to place 1 drop of water on the nickel. Respond to the following:

4. Predict how many drops of water will fit on the nickel before it spills over the edge.

Step 7: Count the number of drops that you add before the water spills over the edge of the nickel. Respond to the following:

5. How many drops fit on the nickel? (Don't forget the drop added in step 6.)

Step 8: Repeat both experiments to see if you can beat your original scores. Respond to the following:

6. How many drops fit on the coins this time?

Penny _____

Nickel _____

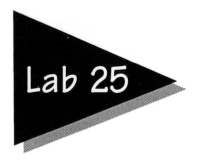

PENDULUMS AND BOB WEIGHT

Lab 25

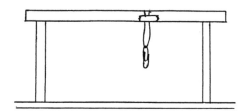

Recorder _____

Controller _____

Hopper _____

Observer _____

Purpose

To study how changes in bob weight affect the frequency of a pendulum

Materials

spool of thread
scissors
10 paper clips
tape
clock with a second hand

Procedure

Step 1: Cut a piece of thread the length of this page. Fold it in half to form a loop and carefully tape it to the vertical edge of your work table. The edge of the tape must be even with the bottom edge of the table top. (See drawing above.)

Step 2: Hang 1 paper clip from this loop of thread to form a pendulum. Swing the pendulum to be sure that it will swing freely without hitting anything.

Step 3: Before continuing, review these two important terms: A *cycle* is a complete swing out and back to its starting position. The *frequency* of a pendulum is the number of cycles the pendulum makes in one minute.

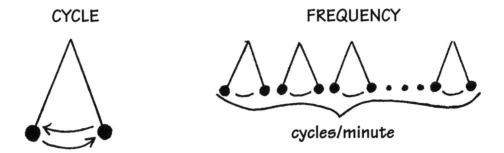

CYCLE   FREQUENCY

cycles/minute

Step 4: When the observer is ready to watch the clock for 1 minute, start counting the cycles of the pendulum. Record the number of cycles on the data chart beside 1 clip under test 1. That is your pendulum's frequency. Label all answers c/m (cycles per minute). Repeat this step for tests 2 and 3.

Step 5: Make your pendulum heavier by hanging 10 paper clips on the loop of thread. The observer will again watch the clock for 1 minute while the controller counts the cycles. Record the frequency on the chart beside 10 clips under test 1. Repeat this step for tests 2 and 3.

Step 6: Now connect the 10 paper clips together to form a chain and hang them on the looped thread. Count the cycles for 1 minute and record the frequency for this trial on the chart beside 10-clip chain under test 1. Repeat this step for tests 2 and 3.

Respond to the following:

1. Did the arrangement of the 10 paper clips affect the frequency of the pendulum? If so, explain how.

2. How could you decrease (slow down) the frequency of a pendulum? Refer to your data chart if you need help. How could you increase (speed up) the frequency of a pendulum?

Pendulums and Bob Weight Data Chart

	TEST #1	TEST #2	TEST #3
1 CLIP			
10 CLIPS			
10 CLIP CHAIN			

Team Science © 1996 Zephyr Press, Tucson, AZ

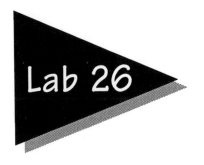

Lab 26

PENDULUMS AND AMPLITUDE

Recorder   _____

Controller   _____

Hopper   _____

Observer   _____

Purpose

To study how changes in amplitude affect the frequency of a pendulum

Materials

spool of thread
scissors
10 paper clips
tape
clock with second hand

Procedure

Step 1: Cut a piece of thread twice the length of this page. Fold it in half to form a loop and carefully tape it to the vertical edge of your work table. (See the drawing above.)

Step 2: Open the outside curve of a paper clip to make a hook and hang it on the loop of thread. Hang 9 paper clips on the hook to make a heavy pendulum bob. It should swing freely without hitting anything.

Step 3: Before continuing, review these three important terms: A *cycle* is a complete swing out and back to its starting position. The *frequency* of a pendulum is the number of cycles it makes in one minute. *Amplitude* is the size of the swing.

Step 4: Test the frequency for large amplitude first. Keeping the thread taut, pull the bob weight far to one side. When the observer is ready to watch the clock for 1 minute, release the bob weight and begin counting the cycles. At the end of 1 minute record the number of cycles by large amplitude under test 1. Do this test twice more taking care to release the bob weight from the same starting place each time. Record the frequencies for tests 2 and 3. Label all answers with c/m (cycles/minute).

Step 5: Measure the frequency for medium amplitude by pulling the bob weight half as far to the side as you did for large amplitude. Release the bob weight and begin counting when the recorder is ready to watch the clock for 1 minute. Record the number of cycles by medium amplitude under test 1. Repeat the test, taking care to release the bob weight from the same starting place each time. Record the frequencies for tests 2 and 3.

Step 6: Repeat the procedure outlined for large amplitude and medium amplitude, but this time pull the bob weight only a small distance to the side. As before, count the number of cycles as the recorder watches the clock for 1 minute. Record the frequency of small amplitude for tests 1, 2, and 3. Respond to the following:

1. How does the amplitude of a pendulum affect the frequency?

Pendulums and Amplitude Data Chart

	TEST #1	TEST #2	TEST #3
LARGE Amplitude			
MEDIUM Amplitude			
VERY SMALL Amplitude			

Team Science © 1996 Zephyr Press, Tucson, AZ

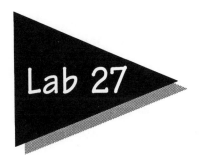

Lab 27

PENDULUM SWINGS

Recorder _____

Controller _____

Hopper _____

Observer _____

Purpose

To learn how to push a swinging pendulum so that the energy is transferred to its natural frequency

Materials

spool of thread
scissors
tape
4 large paper clips
2 pennies
1 soda straw

Procedure

Step 1: Securely tape 2 paper clips 1/2 straw length apart to the top of your work table so they extend over the edge at least 1/2 their length.

Step 2: Hang two paper clips, one from each of the first two. Be sure these two clips can move freely in the paper clips taped to the work table. Thread a straw through the bottoms.

Step 3: Hang a penny from each end of the straw using tape and thread. (See drawing on page 67.) Make one penny pendulum about twice as long as the other.

Step 4: Push the center of the straw to make the pendulums swing. Practice pushing the straw until you know how to add energy to the natural frequency of each pendulum. Respond to the following:

1. When do you push the straw to add energy to the short pendulum?

2. When do you push the straw to add energy to the long pendulum?

Step 5: Change the pendulums so that they have the same natural frequency and you can push the straw to add energy to both at the same time. Respond to the following:

3. How did you change the pendulums so both had the same frequency and would accept energy at the same time? Draw how they looked.

4. Why can't energy be added to the long and short pendulums at the same time?

Team Science © 1996 Zephyr Press, Tucson, AZ

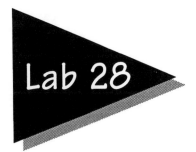

Lab 28

OWL PELLETS

Recorder _____

Controller _____

Hopper _____

Observer _____

Purpose

To study the eating habits of owls by examining the skeletal remains found in an owl pellet; to clean, sort, and classify bones before reconstructing a skeleton.

Materials

owl pellet

tweezers

toothpicks

egg carton

balance scale

ruler

paper plate

5-by-7-inch card

white glue

hand lens

sheets of paper, 6 square cm

Procedure

Step 1: Place a sheet of the paper on the left side of the balance scale. Set the slider at 0 and balance the scale using the adjustment knob. This procedure will ensure that you don't include the weight of the paper in the weight of the owl pellet.

Step 2: Remove the foil cover before placing the owl pellet on the small paper to weigh it. The pellet has been fumigated to kill any bacteria or parasites and the foil is simply a protective covering.

Measure its length and width with the ruler. Record these below:

Weight: _____

Length: _____

Width: _____

Step 3: Place the owl pellet on the plate and use the hand lens to examine it. Although the pellets are glossy black when they are regurgitated, their appearance changes somewhat as they dry. Respond to the following:

1. Describe the color, texture, and any other interesting features that are visible through the hand lens.

2. Copy the grid below onto your paper and make a scale drawing of your pellet by outlining the pellet on the grid and adding any visible surface details.

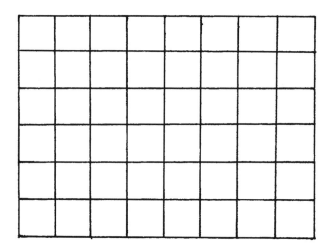

Step 4: Now you are ready to see what your owl pellet contains. Begin by gently pulling the pellet apart with the tweezers and the toothpicks. Separate the bones from the fur and feathers. Carefully clean the bones before sorting them into the egg carton. Thoroughly clean the skulls or jaw bones because they will be most useful for identification. Respond to the following:

3. How many skulls or pairs of jawbones are in your owl pellet?

4. How many are from mammals? How many from birds? (See page 73.)

5. Assuming that the owl produces 2 pellets per day, how many prey animals would the owl eat in one day? In one month? In one year?

Team Science © 1996 Zephyr Press, Tucson, AZ

Step 5:   Identify the skulls or jawbones found in your owl pellet by matching them with the skulls on the Study Key (see p. 72) Then respond to the following:

6.   List the prey animals found in your owl pellet and tell how many there are of each kind.

Step 6:   Refer to the Study Key to select the bones of a vole from your egg carton. Arrange the bones in the proper order on the 5-by-7-inch card. When your prey animal is complete, lightly apply glue to each bone with a toothpick and assemble the animal. Although most owl pellets contain at least one vole, reconstruct another prey animal from your pellet, if necessary. Then respond to the following:

7.   What important information can be obtained from owl pellets?

Owl Prey Characteristics

Old World rats and mice
Supraorbital ridge (faint in mice); molars with tubercles in 3 longitudinal rows

Pocket gophers
1 to 2 vertical grooves on upper incisors

Shrews
Teeth tipped with red

Voles
Molars irregular triangles in rows more than 5 mm long

White-footed mice
Molars with tubercles or flat crowned in 2 longitudinal rows less than 4 mm long

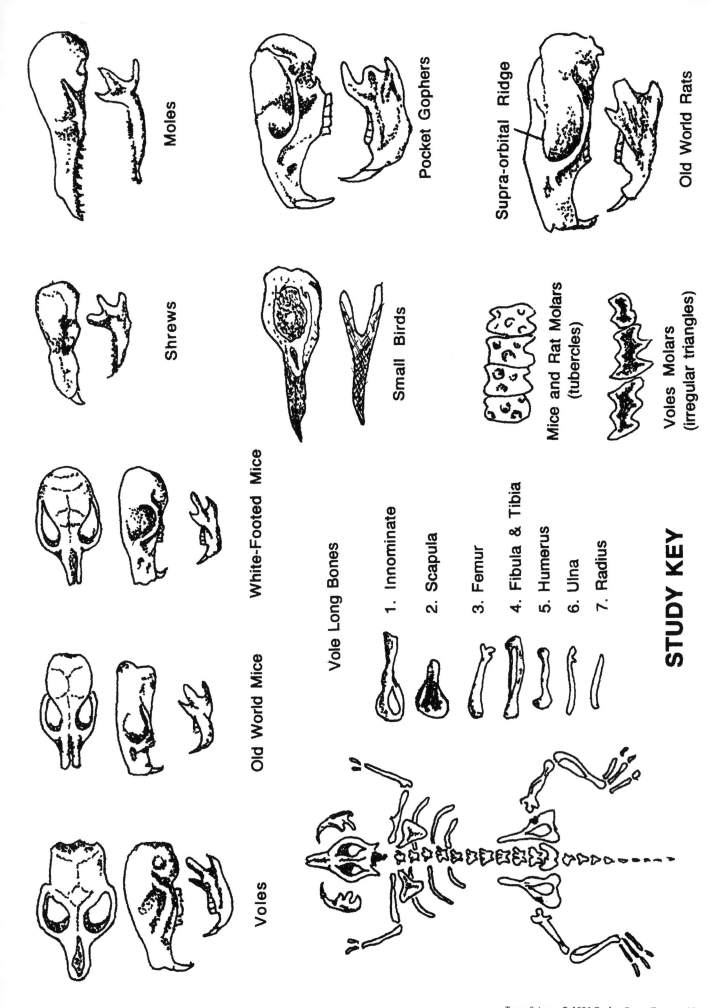

Moles

Pocket Gophers

Supra-orbital Ridge

Old World Rats

Shrews

Small Birds

Mice and Rat Molars
(tubercles)

Voles Molars
(irregular triangles)

White-Footed Mice

Old World Mice

Voles

Vole Long Bones

1. Innominate
2. Scapula
3. Femur
4. Fibula & Tibia
5. Humerus
6. Ulna
7. Radius

STUDY KEY

SKULL FLOW CHART

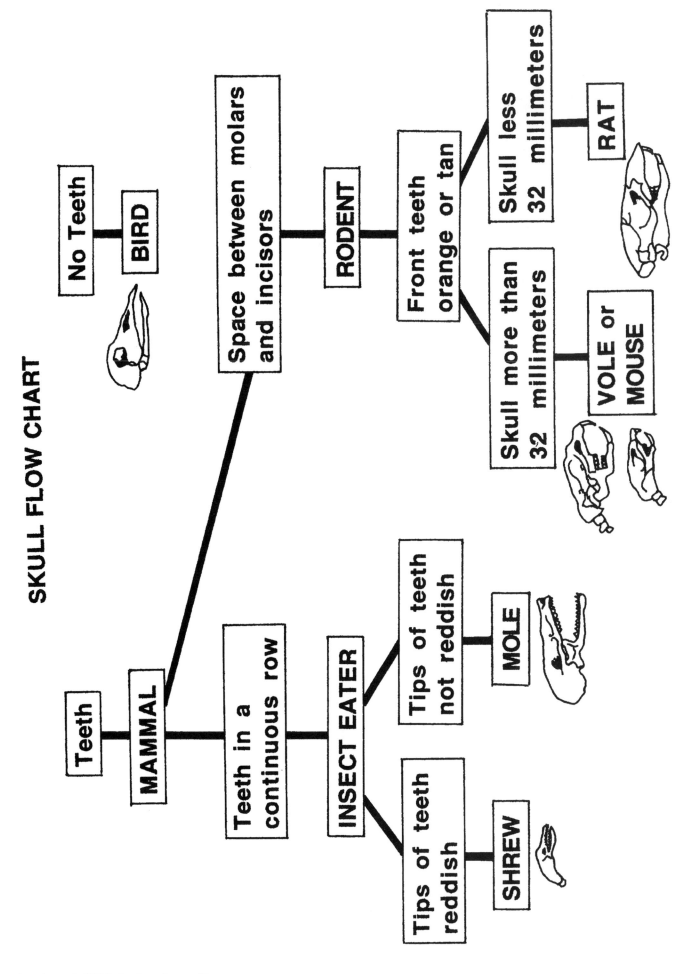

Teeth → MAMMAL

No Teeth → BIRD

Space between molars and incisors → RODENT

Front teeth orange or tan

Skull less than 32 millimeters → RAT

Skull more than 32 millimeters → VOLE or MOUSE

Teeth in a continuous row → INSECT EATER

Tips of teeth not reddish → MOLE

Tips of teeth reddish → SHREW

Team Science © 1996 Zephyr Press, Tucson, AZ

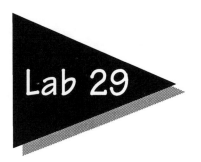

DISTANCES IN THE SOLAR SYSTEM

Lab 29

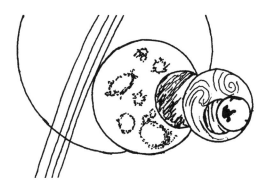

Recorder _____

Controller _____

Hopper _____

Observer _____

Purpose

To make a scale model showing the relative distances between the planets in the solar system and their relationships to the sun

Materials

meter stick
strip of paper 1 m x 10 cm
tape
crayons

Data: Mean distance from the sun

Mercury:	57,900,000 km
Venus:	108,200,000 km
Earth:	149,600,000 km
Mars:	228,000,000 km
Jupiter:	778,400,000 km
Saturn:	1,424,600,000 km
Uranus:	2,866,900,000 km
Neptune:	4,486,100,000 km
Pluto:	5,890,000,000 km

Team Science © 1996 Zephyr Press, Tucson, AZ

Procedure

Step 1: Tape the paper strip to your work table, and color a large yellow sun at one end.

Step 2: Use the scale 10,000,000 km = 1 mm for your model. Scale down the distances to fit onto the paper strip by dividing each distance by 10,000,000. An easy way to do that is to move the decimal point 7 places to the left and round off to the nearest whole millimeter. Determine the scaled-down distance between each planet and the sun. Respond to the following:

    1. Write the new scaled-down distances for each planet.

Step 3: Mark the location of each planet on your paper strip by measuring its relative distance from the sun. For example, Mercury's scaled-down distance is 5.79 mm, which rounds off to about 6 mm. Measure 6 mm from the sun, mark the spot, and label it.

Step 4: Continue measuring and labeling clearly each planet; add recognizable features (that is rings, moons) where appropriate.

    2. Are the relative distances greater between the inner planets or the outer planets?

    3. Between which two neighboring planets is the greatest gap in space?

    4. Which two planets are closest to each other?

    5. According to the scale you used, how many kilometers are represented by 1 millimeter?

Step 4: When you have completed your scale model, write all group members' names in the bottom right corner and turn it in to the teacher.

Lab 30   EARTHQUAKE DRILL

Recorder   _____

Controller   _____

Hopper   _____

Observer   _____

Purpose

To demonstrate knowledge about earthquakes and preparation for earthquakes; to predict possible damage and resulting problems; to find solutions for the problems caused by earthquakes

Materials

map
paper
pencil

Data

Place: Anaheim, California
Date: Tuesday, May 5, 19__ (Cinco de Mayo)
Time: 12:35 P.M.
Magnitude: 8.5 on the Richter Scale

Procedure

Step 1: Study the map on page 78 until you are thoroughly familiar with it. This is very important! For the purpose of this lab you are at the location assigned in the following list when the earthquake occurs.

Lab 1: Madison Junior High School
Lab 2: IBM Factory (electronics manufacturer)
Lab 3: Anaheim General Hospital
Lab 4: Mobil Gas station
Lab 5: Bob's Big Boy Restaurant
Lab 6: Angel Stadium
Lab 7: Ralph's Supermarket
Lab 8: Large intersection (parade in progress)

Step 2: The controller will act as the chairperson. The observer will present your findings to the rest of the class. You may use visual aids: pictures, charts, graphs, diagrams, props, and so on. Other members of the group may assist the observer in the presentation.

Step 3: Brainstorm to make a list of the possible damage and the problems that could arise at your location as a result of an earthquake of this magnitude. For each problem, find a solution that would minimize the danger. Be sure to include the following in your plan:

- Transportation home
- Medical care and treatment for injured people
- Measures to prevent further damage from aftershocks

Step 4: Fold a piece of notebook paper into three columns. Write *Damage* at the top of column 1, *Problems* at the top of column 2, and *Solutions* at the top of column 3. This sheet will help you organize your notes for your oral presentation. List the type of damage that you would expect from an earthquake with a magnitude of 8.5. Next to each item, list the possible problems caused by that damage, and finally include at least one possible solution for each problem.

Step 5: Tomorrow each group will have 3 minutes to describe the damage at their location, list the problems caused by that damage, and explain how they would solve those problems. There will be an additional 3 minutes to answer questions.

TEACHER DIRECTIONS

Contact your local Red Cross. They have free pamphlets for first aid and earthquake preparedness as well as videos and speakers that will prepare students for this lab.

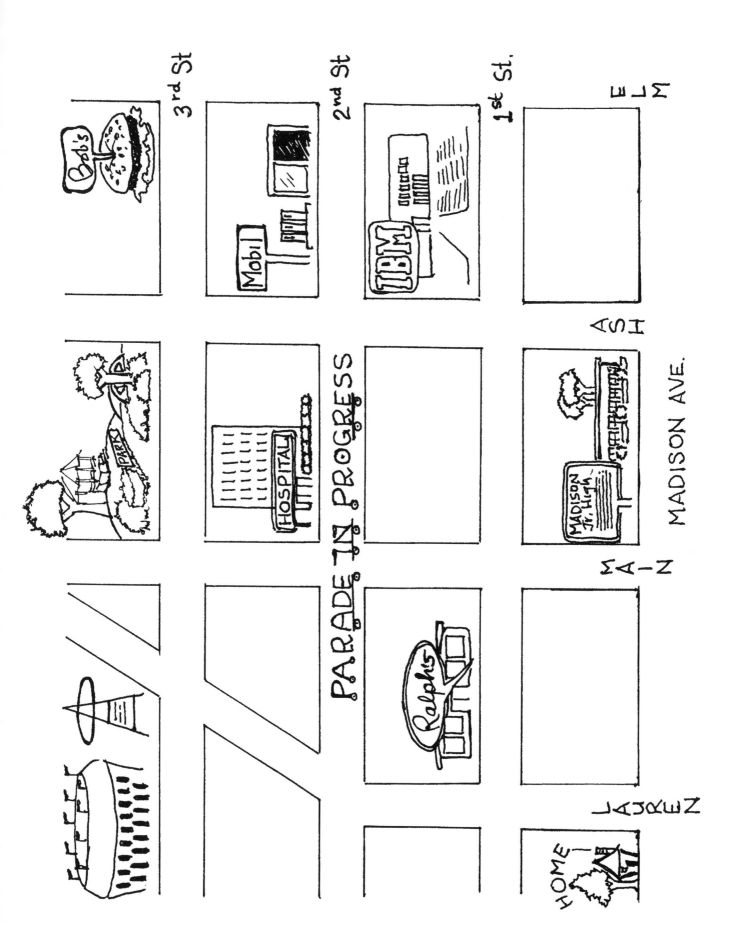

Lab 31 JUICY FRUIT

Recorder _____

Controller _____

Hopper _____

Observer _____

Purpose

To measure and compare the amount of water in various kinds of fruit

Materials

balance scale
5 paper plates
5 different kinds of fresh fruit

Procedure

Step 1: Label each plate with the name of a fruit.

Step 2: Before weighing the fruit, adjust the balance of the scale to 0, if necessary.

Step 3: Weigh each fruit on its paper plate. Record the weight to the nearest tenth of a gram on the data chart. Replace the numbers with the names of the fruit used. Respond to the following:

   Hypothesis: Which fruit do you think contains the greatest percentage of water?

Step 4: Cut each fruit into tiny pieces. Hold the fruit over the plate as you cut so that all of the pieces remain on the plate. The smaller the pieces, the faster dehydration will occur.

Step 5: Leave the plates uncovered and place them in a warm, dry place where they will not be disturbed.

Step 6: Each day, weigh the plates of fruit and record the weights on the data chart. Continue to weigh each plate of fruit until it stops losing weight. Some fruit will take longer than others.

Step 7: When the fruit has stopped losing weight, it is time to calculate the percentage of water in each fruit. First find the weight of the water each fruit contained by using this formula:

(Weight before drying) - (Weight after drying) = Weight of water

Use this formula to find the percentage of water in each fruit:

$$\frac{\text{Weight of water}}{\text{Weight before drying x 100}} = \text{Percentage of water in the fruit}$$

Step 8: Record the results of your tests on the chart. Replace the numbers with the names of the fruit used.

- Was your hypothesis correct?

- Discuss any interesting or unusual observations you made while doing this experiment.

- What would you do differently if you were to do this experiment again?

Team Science © 1996 Zephyr Press, Tucson, AZ

Daily Weight Chart

Fruit	Fruit #1	Fruit #2	Fruit #3	Fruit #4	Fruit #5
Day 1					
Day 2					
Day 3					

Percentage of Water Chart

	Fruit #1	Fruit #2	Fruit #3	Fruit #4	Fruit #5
% of H_2O					

Lab 32   AIR QUALITY

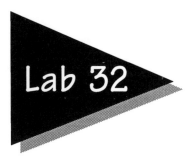

Recorder _____

Controller _____

Hopper _____

Observer _____

Purpose

To study and compare the quality of air in various locations

Materials

4 pieces of cardboard, 4 cm by 20 cm
clear adhesive paper
ruler
pen
spool of thread
hand lens
scissors

Procedure

Part 1

Step 1: Mark off 2-cm squares on one side of each piece of cardboard and number them. The ruled grid will help you count the number of particles collected on your air pollution sampler.

Step 2: Staple about 60 cm of thread to one end of each sampler. Tie the thread around the staple so it won't slip.

Team Science © 1996 Zephyr Press, Tucson, AZ

Step 3: Cut 4 pieces of clear adhesive paper to fit the cardboard samplers. Without removing the backing from the pieces of paper, staple them to the ruled side of each piece of cardboard with the backing on the outside. Once your sampler is at the selected location, remove the backing to reveal the sticky surface. Your numbered grid should be clearly visible through the Contact paper.

Step 4: Hang the air-quality samplers at selected locations where they will not be disturbed for one week. You may choose any place that interests you—cafeteria, smoking and nonsmoking sections in a restaurant, your home, your backyard, supermarket, cleaners, car, and so on. Be creative.

Step 5: The samplers are due back in class at the end of one week. It is best to carry them inside clean boxes with lids so that nothing can disturb the particles you collected during the week.

Part 2

Step 1: Using a hand lens, count the number of particles in any 10 selected squares and record the information on the data sheet on page 84. Include any other interesting observations.

Step 2: Add up the number of particles found in the 10 squares and divide the total by 10 to find the average. You can compare this average to compare to other locations.

Step 3: Prepare your findings to present to the rest of the class. Describe your location and tell the average number of particles found in the squares. Be ready to discuss what the particles might be and what might be done to reduce the number of particles.

Air Quality Data Sheet

Square	Number of Particles	Observations

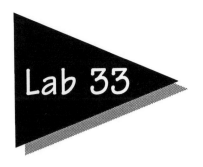

Lab 33

BIODEGRADABLE MATERIALS?

BIODEGRA...

Recorder	_____
Controller	_____
Hopper	_____
Observer	_____

Purpose

To examine the biodegradability of various materials in salt water, fresh water, a dry landfill, an active landfill, and under the sun

Materials

5 newspaper pages

5 brown paper bags

10 biodegradable plastic bags (2 brands)

10 nonbiodegradable plastic bags (2 brands)

8 one-gallon plastic milk bottles

string

1 net

tap water

salt water (15%)

20 ft wire fencing

4 wooden posts

leaf pile (3 ft high)

compost pile (5 ft x 5 ft x 4 ft): compost starter culture, fertilizer, rotting vegetation, grass clippings, and leaves

poster board

Procedure

Step 1: Divide the newspaper pages, brown paper bags, and plastic bags into 5 equal sets.

Step 2: Measure off a 5-foot-square area for the mulch pile (simulated active landfill). Stretch the wire fencing around the 4 wooden posts placed at the corners of the square. Fill the area with 4 ft of grass clippings, leaves, rotting vegetation, fertilizer, and compost starter. Place one set of newspaper and bags in the middle of the mulch pile and wet thoroughly. Continue to wet at regular intervals.

Step 3: Place a second set of newspaper and bags in the middle of a 3-ft-high pile of leaves (dry landfill).

Step 4: Tie a third set of newspaper and bags onto a net that is stretched out in the sun.

Step 5: Cut the tops off the plastic milk bottles. Divide the items in the fourth set of newspaper and bags among 4 of the bottles and fill the bottles with tap water (lake). Add more water as needed.

Step 6: Divide the fifth set of newspaper and bags among the other 4 bottles and fill the bottles with 15 percent salt water (ocean). Add more salt water as needed.

Step 7: Allow all materials to remain in their environments for at least 3 months.

Step 8: Remove each set of newspaper and bags from their environment and place in containers labeled *direct sun, active landfill, dry landfill, lake,* and *ocean.* Identify each item with a felt marker to prevent possible mix-up.

Step 9: Examine each item and record on your data sheet the changes caused by the different environments. Respond to the following:

    1. How long were the materials in their environments?

    2. Which material decomposed the most?

    3. Which material decomposed the least?

    4. Which environment caused the most decomposition?

    5. Which environment caused the least decomposition?

    6. Could you tell any difference between the biodegradable plastic bags and the nonbiodegradable plastic bags?

Step 10: Prepare a large poster board to display the samples from each environment.

Team Science © 1996 Zephyr Press, Tucson, AZ

Biodegradability Data Sheet

Materials	Direct Sun	Active Landfill	Dry Landfill	Lake	Ocean
Newspaper					
Brown bags					
Bio-degradable plastic bags					
Non-bio-degradable plastic bags					
Plastic milk bottles					

Team Science © 1996 Zephyr Press, Tucson, AZ

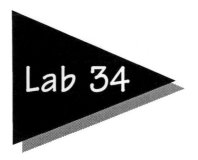

Lab 34 ▶ PASSIVE SOLAR HEAT IS NEAT

Recorder _____

Controller _____

Hopper _____

Observer _____

Purpose

To see which materials radiate heat the longest, making them the best passive solar heat source for a home

Materials

4 three-liter bottles
medium-size stones
flat black paint
soil
4 long thermometers
water
masking tape

Procedure

Step 1: Paint the 4 bottles black up to the neck. Leave the top 8 cm clear so you can see the thermometer. Allow the paint to dry.

Team Science © 1996 Zephyr Press, Tucson, AZ

Step 2: Fill one bottle about half full of soil. Position a long thermometer so that the bulb end will be in the middle of the bottle. Hold it in place while you continue to fill only to the top of the paint. Do not move the thermometer once the soil is in place or the thermometer will break.

Step 3: Seal the top of the bottle with masking tape by completely taping the top around the thermometer. Obscure as little of the thermometer as possible.

Step 4: Repeat steps 2 and 3 as you fill the remaining bottles with stones, water, and air.

Step 5: Label the bottles *soil, stones, water,* and *air.* Respond to the following:

  1. Which material do you think will acquire the most heat during the day? The least heat?

  2. Which material do you think will radiate heat for the longest period of time?

Step 6: Place the four bottles in a window that receives a large amount of sun. Record the temperature of each bottle just before sunset. Continue recording the temperature of each bottle every hour after sunset until there is little temperature change.

Step 7: Collect the data each evening for several weeks. Make a chart to record the temperature for each bottle. Respond to the following:

  3. After analyzing your data, note which material acquired the most heat during the day. Note which material acquired the least heat.

  4. Which material would make the best passive solar heater for the home because it radiated heat the longest?

  5. Is the best passive solar heater the one that acquires the most heat or the one that radiates the heat for the longest period of time?

APPENDIX A

FORMS

Name: _____

Period: _____

SCIENCE EFFORT CREDIT LIST

Below is a required list of activities that will determine your effort grade in science.

Ten activities are due each quarter (5 student choices and 5 teacher choices). You may turn in 2 the first week of the quarter, but only one each of the remaining weeks. All activities must be turned in by Friday (even if you are absent) or they will count for the following week. Each activity is worth 20 points and may be done only once per quarter. No more than 10 effort credits will be accepted per quarter. Keep this list in your science notebook so you can monitor your progress.

Student Choice (Choose 5)

_____ 1. Book report (scientific topic)

_____ 2. Magazine article or pamphlet (scientific topic)

_____ 3. Newspaper article (scientific topic)

_____ 4. Scientific drawing (internal organs of plant or animal)

_____ 5. TV program or video (scientific topic)

_____ 6. 50 aluminum cans (clean, dry and crushed)

Teacher Choice Activities

Teacher Choice

_____ 1. _____

_____ 2. _____

_____ 3. _____

_____ 4. _____

_____ 5. _____

Effort Grade

A+ = 200 points

A = 180 points

B = 160 points

C = 140 points

D = 120 points

Student Signature

_____ Date _____

Parent Signature

_____ Date _____

Team Science © 1996 Zephyr Press, Tucson, AZ

SCIENTIFIC BOOK REPORT

Title _____

Author _____ Publication Date _____

TURN IN THE BOOK WITH THIS PAPER. Write 10 scientific facts that you learned from the book. Begin each fact on a new line, and number the facts. Neatness, spelling, and punctuation count.

_____

_____

_____

_____

_____

_____

_____

_____

_____

_____

_____

_____

Name _____

Date _____

Period _____

SCIENTIFIC MAGAZINE ARTICLE REPORT

Magazine _____

Title of Article _____ **Publication Date** _____

TURN IN THE MAGAZINE WITH THIS PAPER. Write 10 scientific facts you learned from the magazine article. Begin each fact on a new line and number it. Neatness, spelling, and punctuation count.

_____

_____

_____

_____

_____

_____

_____

_____

_____

_____

_____

_____

SCIENTIFIC NEWSPAPER ARTICLE REPORT

Newspaper _____

Name of Article_____ **Publication Date**_____

TURN IN THE ARTICLE WITH THIS PAPER. Write 10 scientific facts that you learned from this article. Begin each fact on a new line and number it. Neatness, spelling, and punctuation count.

_____

_____

_____

_____

_____

_____

_____

_____

_____

_____

_____

SCIENTIFIC DRAWING

DIRECTIONS

1. Use the back of this paper for the drawing.

2. Draw the inside of a plant or animal.

3. Do not trace the drawing.

4. The drawing must be outlined in ink, labeled in ink, and colored with at least 8 colors.

5. The drawing must have a title.

6. The drawing must have at least 16 labeled parts.

7. Print all labels horizontally, and use a ruler to draw the lines connecting the labels to the parts.

8. Neatness and spelling count.

9. Turn in the original with your drawing.

Sit Back and Tune In!

Name _____

Date _____

Period _____

SCIENTIFIC VIDEO OR TV SHOW

Title _____

Channel _____ Date _____ Time _____ (TV show only)

IF YOU VIEWED A VIDEO, TURN IT IN WITH THIS PAPER. Write 10 scientific facts that you learned from this program. Begin each fact on a new line and number it. Neatness, spelling, and punctuation count.

_____

_____

_____

_____

_____

_____

_____

_____

_____

_____

_____

_____

50 ALUMINUM CANS

DIRECTIONS

1. Fifty aluminum cans must be **CLEAN**, **DRY**, and **CRUSHED**.

2. Fill out this paper and attach it to the outside of your bag of cans.

3. Write your name on a piece of paper *inside* the bag as well.

4. Place your bag of cans in the container.

GROUP GRADE SHEET

Staple Group Grade Sheet to the front of lab folders. Monitor cooperative group skills during the lab by circling numbers from 5 to 100 as the lab demonstrates positive group behavior. The group receives the last number circled as their behavior score for the lab.

Group: _____

Period: _____

100 95 90 85 80 75 70 65 60 55 50 45 40 35
30 25 20 15 10 5 0

Date: _____

1. _____

2. _____

3. _____

4. _____

100 95 90 85 80 75 70 65 60 55 50 45 40 35
30 25 20 15 10 5 0

Date: _____

1. _____

2. _____

3. _____

4. _____

100 95 90 85 80 75 70 65 60 55 50 45 40 35
30 25 20 15 10 5 0

Date: _____

1. _____

2. _____

3. _____

4. _____

100 95 90 85 80 75 70 65 60 55 50 45 40 35
30 25 20 15 10 5 0

Date: _____

1. _____

2. _____

3. _____

4. _____

100 95 90 85 80 75 70 65 60 55 50 45 40 35
30 25 20 15 10 5 0

Date: _____

1. _____

2. _____

3. _____

4. _____

APPENDIX B
SUPPLIERS

Call or Write for Catalogues

Burminco (minerals)
128 S. Encinitas
Monrovia, CA 91016
818-358-4478

Carolina Biological Supply Co.
Burlington, NC 27215

**Connecticut Valley Biological
Supply Co.**
Valley Road
Southhampton, MA 01703

DAMON/Educational Division
115 Fourth Ave.
Needham, MA 02194

Edmund Scientific Co.
103 Gloucester Pike
Barrington, NJ 08007

Fisher Scientific Co.
Stansi Educational Materials Div.
1259 Wood Street
Chicago, IL 60622

Frey Scientific Co.
905 Hickory Lane
P.O. Box 8101
Mansfield, Ohio 44905
1-800-225-FREY

Hubbard Scientific Co.
2855 Shermer Road
Northbrook, IL 60062

LaPine Scientific Co.
6001 Knox Ave.
Chicago, IL 60018

Nasco West
1524 Princeton Ave.
P.O. Box 3837
Modesto, CA 95352-3837
414-563-2446 (West)
209-529-6957 (East)

Sargeant-Welch Scientific Co.
7300 N. Linder Ave.
Skokie, IL 60077-3892

Science Kit, Inc.
777 E. Park Dr.
Tonawanda, NY 14150
1-800-828-7777

Scientific Glass Apparatus Co.
737 Broad St.
Bloomfield, NJ 07003

Tri-Ess Sciences, Inc.
1020 W. Chestnut St.
Burbank, CA 91506
213-245-7685
818-247-6910

Take mundane rote learning out of your reviews and make it fun with loops for groups...

THINK MATH!
Interactive Loops for Groups
by Dale Bulla
Grades 4–10

Invite your students to participate in loops for groups and you'll have their rapt attention.

Look to *Think Math!* for learning strategies that are quick and easy to bring to your classroom. Use interactive loops to add excitement and challenge to reviewing material and practicing skills. You'll find plenty of reproducible samples to get you started.

176 pages, 8 1/2" x 11", spiral.
1064-W . . . $29

Use intrigue to involve students in learning

GUMSHOE GEOGRAPHY
Exploring the Cultural, Physical, Sociological, and Biological Characteristics of Our Planet
by Richard S. Jones
Grades 6–12

Intrigue students with a game format that encourages them to learn about our planet. Select from 176 fun, reproducible activities that you can integrate into your social studies, history, language arts, and math units.

You'll be impressed by the depth and breadth of the activities. And each one includes a teacher guide—making it easy for you to teach geography.

320 pages, 8½" x 11", spiral.
1067-W . . . $39

CALL, WRITE, OR FAX FOR YOUR FREE CATALOG!